baby WALRUSES

KIM THOMPSON

CREATIVE EDUCATION • CREATIVE PAPERBACKS

CONT

CONTENTS

I AM A CALF.

I am a baby walrus.

I do not have tusks yet. They will grow later.
nose
back flipper
front flipper

I was born on ice in the Arctic Ocean.
I could swim as soon as I was born.

I drink my mom's milk. It has lots of fat. It helps me grow thick blubber to keep me warm.

My <u>sensitive</u> whiskers help

me find clams on the ocean floor. After 10 minutes, I must come up to breathe.

On land, I stay
with my herd.

By the time I am two, I have tusks.

I use my tusks to break breathing holes in the ice. I use them to pull my big body out of the water.

I will grow big!
I may weigh
more than
a car.

SPEAK AND
LISTEN
GO

OORT!

Can you speak like a calf?
Baby walruses call and grunt.

Listen to these sounds:

https://www.youtube.com/
watch?v=3iH4PMKalVE&t=5s

Now it is
your turn!

WALRUS WORDS

blubber: a thick layer of fat that helps a walrus stay warm in cold waters

herd: a group of animals that live and travel together

sensitive: able to detect very small changes in the environment

tusks: a walrus's canine teeth that grow long and pointed

READING CORNER

Abery, Julie. *Little Walrus (Little Animal Friends)*. Mankato, Minn.: Amicus, 2023.

Lawler, Janet. *Walrus Song*. Somerville, Mass.: Candlewick Press, 2021.

Szymanski, Jennifer. *Arctic Animals (National Geographic Readers)*. Washington, D.C.: National Geographic Kids, 2023.

INDEX

PUBLISHED BY CREATIVE EDUCATION AND CREATIVE PAPERBACKS
P.O. Box 227, Mankato, Minnesota 56002
Creative Education and Creative Paperbacks
are imprints of The Creative Company
www.thecreativecompany.us

LIBRARY OF CONGRESS CATALOGING-IN-PUBLICATION DATA
Names: Thompson, Kim, 1970- author.
Title: Baby walruses / Kim Thompson.
Description: Mankato, Minnesota : Creative Education and Creative Paperbacks, [2026] | Series: Starting out | Includes bibliographical references and index. | Audience: Ages 4-7 | Audience: Grades K-1 | Summary: "Introduce beginning readers to the world of baby walruses with this life science starter. Includes photos, a labeled animal diagram, "Make a Noise" section, glossary, and further resources"-- Provided by publisher.
Identifiers: LCCN 2024043261 (print) | LCCN 2024043262 (ebook) | ISBN 9798889897569 (library binding) | ISBN 9781682778425 (paperback) | ISBN 9798889897699 (ebook)
Subjects: LCSH: Walrus--Infancy--Juvenile literature.
Classification: LCC QL737.P62 T46 2026 (print) | LCC QL737.P62 (ebook) | DDC 599.79/91392--dc23/eng/20241226
LC record available at https://lccn.loc.gov/2024043261
LC ebook record available at https://lccn.loc.gov/2024043262

DESIGN AND PRODUCTION
Design by Rhea Magaro
Production by Beeline Media and Design, Inc.
Art direction by Tom Morgan

PHOTOGRAPHS by Alamy Stock Photo/blickwinkel/H. Baesemann, cover, GTW, 8; Dreamstime/Crazy80frog, 14, Ericlefrancais, 11; Getty Images/Daniel Reinhardt/picture alliance, 7, GEORG WENDT, 5; Shutterstock/Bohbeh, 13, Danita Delimont, 10-11, Elena Yakusheva, 12, Emily Westphal Photo, 2-3, Pole 2 Pole Images, 6-7, Potapov Alexander, 4, Vladimir Melnik, 9

Printed in India